You too can grow Bonsai

BONSAI
IN THE TROPICS

Dorothy & Vita Koreshoff

Illustrations
by
Dot & Vita Koreshoff

Photography by Max Candy

Published by Boolarong Publications

First Published in 1988 by Boolarong Press.

National Library of Australia Cataloguing-in-Publication entry
Author: Koreshoff, Dorothy
Title: Bonsai in the tropics
ISBN: 9781922109491 (pbk.)

      1. Bonsai I. Koreshoff, Vita II. Title.
      III. Title: You too can grow Bonsai.

Published by Boolarong Press, Salisbury, Brisbane, Australia.

Printed and bound by Watson Ferguson & Company, Salisbury, Brisbane, Australia.

Contents

Acknowledgements

This book is the third in a series of Bonsai books for beginners, the original book having been adapted to suit specific growing conditions.

Sadly, Vita died before this book was completed, but it was due to his dedication and love of Bonsai that inspired all of his family, and the people with whom he came in contact, that has made Bonsai a way of life for us all.

All the pictured Bonsai are our own. They have been grown entirely in Australia as the result of our own work, time and loving care.

Living in the Hills district of Sydney we have a marvellous climate. Growing happily side by side are plants of tropical origin as well as species from the colder regions.

List of Figures

Introduction

THIS book has been written specifically for people who know absolutely nothing about Bonsai. Nowadays, there are very few who have never heard the name, although it IS surprising how little is really known about its cultivation. We hope to answer the most common questions, and correct the misconceptions. There is no doubt the majority of people are attracted to Bonsai. Very few still feel that they are starved, tortured trees, where as correctly grown Bonsai are blooming with health and vitality.

The most popular misconception is that the cutting of the roots keep the tree small, this is far from the truth. Cutting the roots keeps the tree HEALTHY.

By cutting away approximately ⅓ of the root system, and replacing it with new soil — the tree does not become rootbound, a condition which will weaken any tree.

The roots tend to hold mystery and fear for most people. They ask, in a hesitant fashion: 'Do you have to cut the tap root or something?'

This cutting of the tap root is only necessary if the plant has been grown from seed in the ground. Its purpose is to stabilize the tree. Its fibrous, feeder roots supply the nutriments.

Tap roots from seed sown in a pot usually break into finer roots when they reach the bottom of the container.

Cutting grown plants only produce fibrous roots.

Having overcome the aversion to pruning the roots, we tend to find there is a greater reluctance to prune the top, for, to start with, the novice or gardener is possibly more attracted to the pretty leaves and flowers. If you admire the form and the artistry of Bonsai, you will have to overcome this.

The word NATURAL is controversial when applied to Bonsai. Some think that the more stylized versions are un-natural. All of the styles are copies from nature, the difference is, some of those styles are un-natural to the area or country where you live.

If a seed has fallen into a small depression in a rock containing a small amount of soil, it will germinate, grow at the normal speed for a while, but will slow down and grow very slowly when the roots have filled up this depression. This will eventually develop into a natural Bonsai.

Similarly, the size of the pot governs the size of the tree. When the balance between the root system and the tree is reached, its growth is slowed down considerably. It still grows, but a fraction of its normal growth.

Fig. 1 *Seed*
Fig. 2 *Depression in rock*
Fig. 3 *Natural Bonsai*

The majority of Bonsai books have been written for the temperate zones of the Northern hemisphere, therefore the timing and Genera are not suitable for the tropics. It has long been necessary to have a book dedicated to growing Bonsai in the countries between the tropics of Capricorn and Cancer. Although a large majority of these countries are classified as tropical, they differ greatly both with the timing and duration of rainfall and degree of winter coolness.

The main active Bonsai areas within the zones are:- Hawaii, India, Indonesia, Malayasia, northern Australia, Singapore and Thailand.

The countries classified as wet and dry tropical, experience predominantly dry conditions. There are 12 months of warm to hot temperatures with only one or two months of very heavy rainy periods. One of the problems these places face is overcoming rapid drying due to excessive heat. These conditions prevail in India, Indonesia, northern Australia and Thailand.

Then there is rainy tropical which climatically, is almost the opposite of the above. There is little difference in temperature between winter and summer, and apart from one to two months of dry conditions, the rest of the year will have some rainfall nearly every day. The atmosphere is humid which tends to encourage large lush leaves which can pose some problems with regards to leaf reduction. Special attention will have to be paid to the potting medium to ensure rapid drainage.

The tropics can also vary temperature wise. There can be entirely different conditions between growing Bonsai in the lower and higher altitudes. The latter can experience a winter dormancy, therefore a wider range of traditional plant material can be used.

Generally speaking, trees of the tropics, both indigenous and introduced species tend to have wider spreading crowns. Very rarely does one see the tall conical outlines of the more northerly countries and altitudes. There will be some who will feel more comfortable styling their Bonsai to the umbrella shape they are familiar with. Others become more inspired by, and more in tune with, the classical Japanese Bonsai styles.

VIII

chapter 1

Instant Bonsai

'Instant' is the catch word of today. It reflects the speed of life, and presupposes the shortage of time.

Using this title, we hope to dispel the well established notions that it takes too long, and only the young should start the hobby, and first you must plant your seed, or that they are handed down from previous growers.

If you thought that Bonsai was too difficult, or you have been disappointed by sowing seed under the impression that they would come up as already trained one hundred year old trees; this book will help you.

To start Bonsai from seed is a fascinating experience, but the beginnings are slow, tedious, and for the budding enthusiast — very dull. By starting that way, only very few people will survive the test of patience. Once you realize that to place a tiny plant in a small container its rate of growth is so reduced, that years later it is still a little seedling in a small pot.

In order to turn a young plant into suitable stock — which in turn will be styled into a Bonsai, it should be grown in the open ground, or a large training pot. This will speed up the growing process. Constant pruning will increase the amount of branches, at the same time, thickening the trunk and creating a compact mass to work with.

In the tropics the best source of plant material will be from nurseries, plants dug from gardens or collecting plants with thick trunks from the wild. This will save you a lot of time.

By purchasing a semi-advanced tree, one can create an instant effect in just an hour or two; rootpruning requiring even less time.

If you are a novice to the Art, the result obtained will only be the 'skeleton' of your future Bonsai, but the tree — instantly shaped from semi-advanced stock has become possessed with an inherent beauty. Rough, unrefined and unfinished as it is, nevertheless, its creator has

Fig. 4 *Skeleton of future Bonsai*

given something of himself, and with time and maintenance can one day be proudly called Bonsai.

As you start to improve the shape of the tree by controlling the new growth, you become more fascinated with the art, and as your collection increases, so does your experience.

Although few things obtained instantly are worthwhile, the instant enjoyment, and pride you will feel from your new creation, will, in this case make it all worthwhile if we have convinced you to start NOW.

Fig. 5 *With time and maintenance has become a Bonsai*

chapter 2

Bonsai or potted tree

Growing a plant in a Bonsai container does not automatically turn it into a Bonsai. If you have a tree growing in a pot and have made no attempt to shape it — you have a potted tree. Bonsai are not made from 'special' trees, but are the result of continual shaping and pruning of either hard wooded shrubs or suitable tree species.

Don't worry, at this point, about your lack of knowledge or ideas on how to shape your tree, the hardest and most important requirement is to make the START. So many people say that they wished they had begun years ago. If you start now, your tree will be increasing in age, and as your knowledge improves, adjustments to its shape are always possible.

Fig. 6 *Bonsai*

Fig. 7 *Potted tree*

chapter 3

How to choose your tree

Choosing the right tree at the start can save you years of problems in the future.

This book is aimed at answering your questions and giving you enough knowledge and confidence to get you started by working on semi-advanced nursery stock. This will provide you with something presentable to admire NOW, instead of waiting years for your seeds and cuttings to develop.

Firstly, the variety should appeal to you. Remember, trees react the same in pots as in the ground. For instance, if you do not like to see the trees without their leaves, don't start with a deciduous plant.

If you are allergic to strong perfumes, don't buy Murraya, or Gardenia etc., their scent is just as potent in a Bonsai pot as it is in the ground.

Naturally small varieties should be used if flowers and fruit appeal; i.e. calamansi or cumquat instead of oranges. The reason for this is that unlike leaves, the flowers and fruit do not become miniature. As an example, a camellia bush producing masses of flowers, can, by removing some of the buds, be made to flower fewer but larger quality blooms. This variation in size is within the maximum and minimum of normal.

It is the smaller reduced size of flowers and fruit on Bonsai that lead people to believe that un-natural reduction has been achieved.

Nature has a way of protecting the quality of her species, making sure that they do not deteriorate. This could be the case if confined spaces, drought, lack of nutriment or other causes TRULY cause the fruit, and consequently the seed, to miniaturize.

When conditions are difficult, everything is sacrificed, leaves and excess flowers are shed thus ensuring that at least one NORMAL sized fruit will mature.

chapter 4

How to select suitable stock for Bonsai

Many select trees intended for Bonsai without much thought given to their potential merits as GOOD material.

Some buy a particular variety because it has a few pretty red leaves or flowers, with few or no branches to work with. There is no reason why such a plant cannot eventually become a good Bonsai, but time can be used more rewardingly if your choice is right from the start.

Hopefully you now realize the distinction between Bonsai and an ordinary potted tree. In one sentence, however:— Bonsai is a tree shaped by man to his own or accepted classical style. To shape something, you have to have material to work on. IN the case of a tree — it needs plenty of branches which represent this needed material.

Let us now consider the complete procedure of selection step by step.

Some nurseries have a very uniform and regular stock of plants. At first glance they may all look similar, by taking a closer view, differences become apparent. Following are the points to help you choose the BEST tree.

1. The plant should have a definite trunkline, wider at the base, tapering gradually towards the top. If thick bark is present, this is an added advantage.
2. The lower branches must be strong and healthy, and should start close to the ground, or no more than ⅓ of the intended height of your finished tree.
3. There MUST be plenty of branches on the tree. This will allow you to cut away those that do not fit your design.
4. See if there are well developed, preferably radiating roots present. If none appear visible on the surface, investigate under the soil, for to find them will help to give the illusion of added age to the tree.
5. The leaves should be naturally small, this will give a better proportion to your new 'Bonsai'.

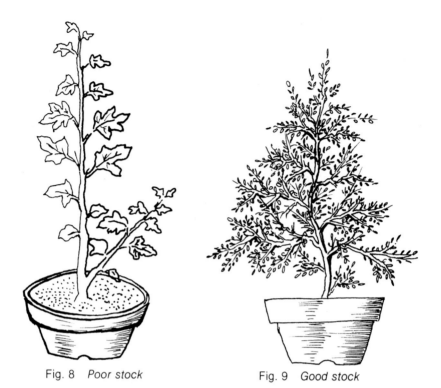

Fig. 8 *Poor stock*   Fig. 9 *Good stock*

Fig. 10 *Radiating root system*

6. The chosen tree should have the ability to shoot from any position on trunk and branches when pruned. The knowledge of trees with such ability is very important for the Bonsai grower as it allows you more scope in the creation of your shaping.

IF AT THE TIME OF CHOOSING YOUR TREE YOU ARE NOT SURE — ASK THE NURSERYMAN, he should be able to tell you if the tree will stand hard cutting back, and if it will sprout on older branches.

6

chapter 5

Trees to avoid

1. Lanky, sparsely branching trees.
2. Too tall with a trunk that has not developed an attractive buttress, and looks like a telegraph pole.
3. Trees with branches too long, with no foliage or branchlets close to the trunk.
4. Branches that start too high on the trunk, are not suitable for the beginner. This type of material is sometimes used by the experienced grower to create Literati style.
5. AVOID TREES THAT ARE NOT HEALTHY. Despite popular belief, Bonsai should not be created from half dead, sick, weak stock. The healthier the plant, the more successful the shock of transplanting and shaping will be. If you are attracted to such a tree, because of its shape — Buy it and plant it into the open ground, or a large training pot for a year or two, until the tree regains its health and vigor.
6. Trees that do not throw new branches on 2 year old wood are hard to manage, and should be avoided until you have gained more advanced techniques.
7. Trees with compound leaves are not a good choice as it is difficult to obtain a 'tree-like' form. Plants such as Calliandra, Poinciana and Jacaranda look pretty when in flower but they are more ferny-looking than tree-like. In horticulture however, there are always exceptions to the rule, and it is possible that a large heavy trunked stump can be transformed into an exceptionally good Bonsai as its compound leaves would be in proportion on a taller tree.
8. Succulent type plants like Jatropha, Brassia and Adenium, which are interesting oddities, are popular because of the heavy buttressed base of trunk. Unfortunately they have large, coarse leaves which do not fit in with the accepted image of Bonsai. The leaves do reduce somewhat, but don't attain a refined appearance. Don't dismiss as unsuitable all large leaved plants, as some do reduce with refinement. Perhaps their worse feature

is a lack of fine branching. These plants though, make attractive companion plants when kept small and as long as they are kept on the dryish side during winter, they are good material for the novice to gain experience. Remember, you should be trying to create a miniature version of a magnificent old tree, and although a thick trunk is one objective, a corresponding branching ramification is also an essential requirement for fulfilling this illusion.

SUMMING UP. After you have successfully created several Bonsai, don't be afraid to work on more difficult stock. To become proficient, the loss of an occasional tree is not too great, but the experience and knowledge gained are.

Casuarina cunninghamiana. Australian Pine, Beefwood, Ironwood, She Oak. Collected from river in 1956. 84cm.

chapter 6

You will need some tools

For your first attempt you will probably have most of the tools required.

You will need:—

- Sharp Scissors.
- Small sharp knife.
- Wire cutters, (electrician's will do).
- Potting stick, (dowel or old ball point pen).
- A piece of fibreglass insect screening. (If unobtainable, double thickness pieces of nylon stocking will do.)
- A metre or two of assorted copper or aluminium wire from (14 gauge to 20 gauge).

Fig. 11 *Sharp scissors*

Fig. 12 *Knife*

Fig. 13 *Wire cutters*

Fig. 14 *Potting stick*

Fig. 15 *Screening*

Fig. 16 *Wire*

Later, as your interest develops, you will find that the range of Japanese imported tools will make the task of cultivating Bonsai easier and more pleasant, as well as more costly.

The most important tool is undoubtedly the Branch cutter. It cuts out the branch, leaving a concave cut. This type of cut has been found to heal more completely than one with a stub left behind from using scissors or secateurs.

For the specialist there are tools ranging from clamps for bending heavy trunks and branches, wire benders, jin (bark) strippers, heavy root pruning sheers to dainty leaf pruning scissors.

Fig. 17 *Branch cutter*

Fig. 18 *Wire cutter*

Fig. 19 *Heavy root pruning shears*

Fig. 20 *Leaf pruning scissors*

chapter 7

When to start

Now that you have obtained the tree and all the necessary materials, the only question remains:— When to start repotting?

'When' is one of the major determining factors for the success or failure of Bonsai survival.

For most varieties, to cut the roots in the middle of summer, or the coldest months of winter will kill them.

It will probably kill most species if the roots are cut during their growing period. There are some exceptions, notably Ficus sp., Pomegranate and Chinese Elm. The best and safest time is when the leaf buds are enlarging, but definitely BEFORE they have opened into new growth. The timing can vary in different areas. If the winter temperatures are low enough to cause the plants to become dormant, then a very safe time is prior to their reawakening in spring. Other areas may experience long, hot, dry periods during which the plants are inactive. At the start of the 'wet' or monsoonal season the plants can be root-pruned as they will soon begin regrowing. If you intend cutting the roots at the beginning of the rainy season, the potting medium MUST be of a sufficiently open granular structure to make it possible for air (oxygen is necessary for root growth) to reach the roots even though the soil is constantly wet.

Another safe time is to catch the period between flowers fading and before any new growth appears. An example is Raphiolepis, the Indian Hawthorn, a broadleaf evergreen tree. Plants that are deciduous i.e. lose all their leaves at one time, are best root-pruned as the leaf-buds are swelling but before they open. Most young plants are root-pruned annually until you have gained the desired structure and dimensions of the framework. After that, less frequent repotting aids in leaf reduction and refinement of the branches.

Figs on the other hand can be repotted at any time during the growing season and as many times as you like. They are very hardy and are excellent plants for the novice.

chapter 8

Choosing a pot

Another common misconception is that the pot must be shallow. This can be true up to a point, certainly most Bonsai pots will be more shallow than nursery containers.

The guidelines for helping you choose a pot are as follows:—

- The thicker the trunk, the deeper the pot.
- Dainty trees with slim trunks, look more attractive in shallower containers.
- Rugged trees giving a masculine feeling, look compatible in rectangular or square pots.
- Graceful styles with curved trunk and branches, suit oval or round pots.
- Cascade definitely need tall pots, while,
- Semi-Cascade, depending on the shape, can go into fairly deep to quite shallow containers.
- Pots that are flat to the ground, are not as healthy for the tree as pots with legs. This allows better drainage and circulation of air around the root-system.
- Unless growing water loving plants such as Willows, Wisterias etc., pots that are glazed inside, do not 'breathe'.
- All pots must have drainage holes.

The main colour range is brown, terra cotta, cream or white, blue and a subtle green.

Conifers usually look better in brown or terra cotta, and deciduous or fruiting and flowering, — any colour that complements its predominant feature.

chapter 9

To prepare the pot

Cover the drainage holes with insect screening that has been cut into squares slightly larger than the openings. Do not cover the entire base with the screening, for as the roots grow, they will entwine in, out and around, making it harder to unravel when next repotting.

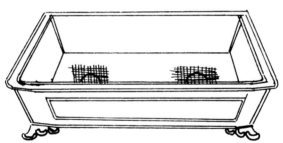

Fig. 21   *Screening placed over drainage holes*

Some people prefer to secure the screening to the pot with wire. This is especially useful when making group plantings, as the placement of trees often dislodges the squares.

Fig. 22   *Securing the screen*

Next, place a layer of potting mix in the bottom of the pot, deep enough to cover the screening, or deeper if necessary to raise the surface roots to the rim of the pot.

A mound of soil should be placed in the position where the root ball will rest. This will assist in filling the area with soil, as air pockets are unhealthy for the roots.

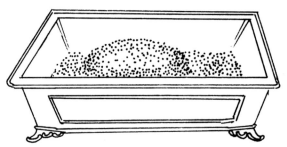

Fig. 23   *Layer of soil and mounded soil in position*

Raphiolepis indica. *Indian Hawthorn. Grown from seed planted in 1949. 45cm.*

14

chapter 10

To prepare the tree

If you are going to style the tree at the same time as its rootpruning, then do so before planting for two reasons:—

1. The tree will be in an unstable condition after potting, making it harder to work also, its not good for the roots to receive so much movement.
2. If styling after planting, you may decide to change the front of the tree.

Water the tree well before cutting the roots. This is contrary to most information given, but we have found that as the tree cannot take up any water through the roots until they have regenerated, the more filled the tree is with water to start with, the better its chances of survival. At the time of shaping, (before cutting the roots) there is an advantage of having the soil in a slightly dry state, for if necessary to make any drastic alteration to the shape by bending, the task will be made so much easier.

If the tree is really weak and in very poor condition, it could be dangerous to either prune the roots or drastically bend and shape the top. The tree MUST be strong and healthy before attempting the operation.

Conifers, especially true Pines (Pinus) should not be stripped of all, or the majority of their foliage, as this procedure considerably weakens the tree. The nutrient solutions that are transported via the roots up the trunk of the tree, to be transformed into sugars, which is the changed form of nutriment acceptable to the tree, cannot be achieved without the foliage present. Thus, a weak defoliated tree may starve to death.

Fertilizing for a season to build up strength, or gradual defoliation is necessary if excessive reduction of foliage is necessary for a particular design.

chapter 11

Soil-potting mixture

Many things can be written about soil and its importance to the plant. Let us start by considering a tree growing in the open ground as compared to a tree grown in a container. The open ground does not restrict the root run, and the tree has the freedom to fossick for nutriments and water at will.

In a container, especially a small Bonsai pot, the confines of the pot restrict such freedom. If it happens to be poor soil which is placed in the container — the roots of the tree will not grow vigorously. If the soil is rich, the roots will grow down and radially very fast, by — passing the central portion of the soil, and reaching its sides and bottom without benefiting from the middle area of the pot.

The roots will congregate on the outside of the rootball getting little opportunity to use the nutriments from the main volume of the soil. The root's volume will increase, thus creating pressure and compacting the soil in the pot to such an extent that it becomes impervious to air and water.

The above is true in a more pronounced degree if using ordinary soil. The pattern of root growth in a container will always be the same no matter what soil or soil mixture we use.

Therefore, the only way to improve matters is to consider how to minimise the gradual, but constant compression of soil in the pot, and to provide better conditions for the tree for longer periods.

Unfortunately soil, defined as a rich fertile top soil containing abundant decaying organic matter, is non-existent in some tropical countries. If this is the case, then equal quantities of sand and humus will do. On the other hand, no matter how well your plants grow in the ground, if the soil is so fine that it will pass through a 1/16th sieve or a piece of insect screening, it will clog the drainage and is better omitted from the mixture. The ideal soil will be granular in form and will not break down easily when wet.

When mixed, the soil should be porous and spongy, and to obtain this condition needs the addition of at least two other ingredients.

To create good drainage is of extreme importance everywhere, but in the tropics it is even more so. Climates that experience constantly wet conditions most of the year would need to have quite large spaces between the solid granules so that air spaces can exist even though there is water passing through.

There is also the tropical dry climate where the emphasis is placed on ensuring that the pots do not dry out and the danger here is that the plants could consequently be over-watered. Over-watering is not the amount you give at any one time, but the frequency of constant rewatering when the soil is still wet. When this happens, and especially if the solid drainage granules are fine, oxygen is excluded from the soil ball. Roots regenerate, and start to grow as the soil is drying out, for they are activated into growth in search of water.

The drainage material will vary according to what is available. A coarse, sharp, washed sand that is retained on a 1/16th inch mesh, about the size of insect screening, is the usual medium. If sand is unavailable, crushed volcanic rock, or scoria is suitable. Crushed brick or tile from stoneware clay is preferable to terracotta crushed tile which absorbs too much moisture and silt. The ideal granular size is from 2 to 5mm or ¼ to 1/16th inch. Another possibility is crushed river stone. This must be sharp so that it aids in the division of roots. Smooth round decorative gravel is not as good as roots will tend to grow in length rather than subdivide into finer rootlets.

Of the various humus possibilities, we prefer to use either horse or cow manure, but whatever manures are available for garden use in your area, should be suitable for Bonsai. One exception is poultry manure. When used as a soil ingredient, because of the amount needed, it is much too strong. When used in moderation as a fertilizer it is excellent. The purpose of the humus, besides supplying food and good bacteria, is to act as a sponge, softening the potting mixture and providing room for expansion of the root-system. The manure you use then should be mild enough to be used to the proportion of ⅓ to ½ depending on the other ingredients used.

If you can buy a commercial compost, or, better still, make it yourself, it will prove to be suitable. Another possibility is to collect leaf mulch from the countryside. This can be just as good. The main requirements of manures and composts is that they be given sufficient time to age. If they have been bought in sealed bags, then they must be emptied and turned several times over a period of 2 to 3 months to allow oxygen to penetrate. As well, do not store the manure in airtight containers, as the beneficial bacteria, which needs oxygen to survive, will be killed. The anaerobic, or bad bacteria that thrives in airless conditions will cause the plant to deteriorate when used in the potting mix.

Peat is the next ingredient and is placed last on the list in our opinion. Its worst feature is that once it becomes slightly dry, and don't

17

forget this is the requirements between waterings, the only way that you can rewet it is to immerse it for more than half an hour.

Make a test for yourself, take two glasses of water and place powdered or milled cow manure on top of one, and peat moss on the other. The manure will instantly absorb the water and sink to the bottom, while the peat will be floating for hours. Peat has no nutritional value, so ⅓ of your potting mix is almost useless.

The proportions to use of each ingredient can vary according to the climatic conditions and type of tree. But as a starting point use by volume equal quantities of sand, soil and humus.

Ficus rubiginosa. Port Jackson Fig. From advanced nursery stock 1949. 62cm.

chapter 12

Root pruning and potting

1. Knock your tree out of the nursery pot by placing your hand over the soil and having the trunk between the first and the second fingers of your hand. Turn the tree upside down and knock the rim of the pot on to the edge of the table. The tree with soil will slip out of the pot very easily.

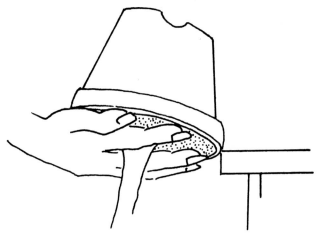

Fig. 24   *Turn upside-down and knock rim of pot against table*

2. Place the tree on the table, and with the sharp end of the potting stick tease about 2½ cm of the soil from the top, or until you reach the main roots growing from the trunk.

Fig. 25 *Tease away until you reach the main root growing from the trunk*

3. Lift and turn the tree upsidedown again and start teasing soil and roots from the bottom of the rootball. Work your stick from the centre of the soilball towards the outside.
4. Do the same from the sides of the rootball. Scraping soil and roots away until you have removed ⅓ to ½ of the old soil. Consider the shape of the Bonsai pot. If it is rectangular, remove more from two sides to make the soil ball to conform to the shape of the pot.

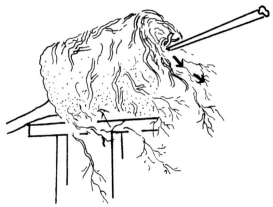

Fig. 26 *Removing soil from sides and bottom*

20

5. Now there are roots exposed around the sides and at the bottom of the rootball. Take the scissors and trim the roots close to the remaining soil. The amount of soil taken away should enable you to place the tree in the pot, with enough room around the sides to fill with your prepared potting mixture.

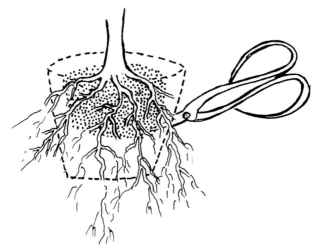

Fig. 27   *Original rootball indicated, exposed roots to be cut away*

6. Now that the tree is placed in the pot giving a pleasing overall look, fill all empty spaces between the old soilball and the walls of the pot. Use a jabbing motion making sure the potting stick reaches right to the bottom of the pot. Add more soil as you proceed. The soil level should be slightly below the rim of the pot to trap the water. The soil used should be dry, allowing it to fall freely between the roots. If the soil is damp, you may have to use your finger to feel for empty spaces, and press in the soil.

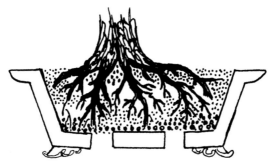

Fig. 28   *Air spaces due to either insufficient potting or damp soil*

chapter 13

Positioning the tree

Without going into much detail, the easiest way would be for you to place the trunk, so that the foliage equally reaches both ends of the pot.

If the shape is assymmetrical ie. one of the lower branches are longer than the other, by planting more to one side of the container, you will find that the foliage should still be equidistance from the edges of the pot.

Fig. 29   *Placing the tree in pot*

chapter 14

Ground covering — moss

Moss is an added attraction to the overall beauty of the tree, and would be the most popular and desirable cover used. It can also be a trap for the novice as the soil being covered becomes rather hard to tell by sight whether it's moist or dry.

Most people think that when the moss is green — the tree should be in a healthy condition.

This is far from the truth, we have seen so many beautiful pots of lush green moss with a dead tree in it . . . The reason being that when moss goes into a dormant condition in summer and becomes a rather dull brown, the owner thinks the plant is not receiving enough water.

By watering to suit the moss, which may be a variety activated by excess moisture, the roots of the tree are suffocating from lack of air due to constant moisture and unless noticed in time, ends in the ultimate death of the tree.

The active growing time for moss is from autumn to spring. It can be lifted at any time and transferred to the top of the soil you wish to cover.

If it is a thick layer, then with a dowel or bottle, can be rolled out to form a thinner layer; also your moss doubles its area. It is healthier for the Bonsai if a thick blanket of moss does not hinder the access of air to the roots.

A heavy coating of moss sometimes restricts the penetration of water to the soil unless the water is applied with some force.

Moss can be collected, allowed to dry, and sieved or broken up and stored until autumn. A few weeks after sprinkling it on top of your Bonsai pot, you will notice a green cover developing.

Watering by soaking up for a week or two will prevent the spores being washed away.

chapter 15

Pebbles

Some people prefer a covering of pebbles, which on some trees are more in keeping with the environment where the plant naturally grows.

For the beginner we would recommend a full year without any covering at all, so that it is possible to more clearly watch the condition of the soil, ie. its condition in regard of water content.

Lagerstroemia Crepe Myrtle. Dug from garden 1966. 29cm.

chapter 16

After care

Immediately after rootpruning, unless transporting your Bonsai some distance, it should be watered thoroughly. The best way would be to immerse it in water over the level of the soil. Some of the humus will float away, but this does not matter. Wait a few minutes until it stops bubbling and remove it. Or, stand it in a tray of water until the moisture reaches the surface. The time will vary according to the depth of water.

If you are repotting at the optimum time for the variety, unless you live in a very dry, exposed and windy area, no special location is necessary. In non humid regions conifers and broad leaved evergreens would appreciate a daily spraying of the foliage, as this is the only way they can absorb moisture until the roots start to grow.

The quicker you can get the soil to become dryish after the initial watering — the better. The roots grow and elongate as the need to search for water becomes necessary, so, the longer the roots remain wet, the slower their regeneration. It is for this reason, that if your area has a definite rainy season, do not repot at that time unless they can be sheltered from constant rain unless the soil is extremely porous.

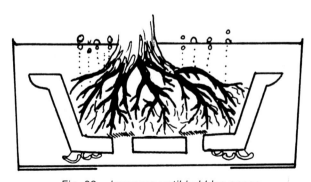

Fig. 30   *Immerse until bubbles cease*

After the initial watering, you can continue its watering program by immersion, or by hose if you have a very fine spray.

Root-pruning out of season will require more sheltered, out of wind, humid conditions. Spraying the foliage several times a day for at least the first week, will help to conserve the moisture content of the plant.

Make sure the humidity level of the sheltered area is not so high as to slow down the drying of the soil.

After two weeks they can be removed to their normal position.

Fig. 31   *Soak until moisture reaches the surface of the soil*

Eugenia smithii. *Lilly Pilly, Brush Cherry. Grown from seed planted in 1966. 65cm.*

26

chapter 17

Maintenance

The most disappointing feature of Bonsai is that they are NOT INDOOR PLANTS. After creating one, it is a great temptation to keep it inside just to look at, or (unfortunately) as an ornament.

Bonsai are trees, and as all trees growing in the wild, need all the NATURAL elements such as rain, dew, wind (in moderation), sun, and variation in temperature from one season to another.

Correct WATERING is one of the most important factors for the well being of your Bonsai. Unfortunately, it can not be on a regular basis, but the pot size, season, and the position will determine how often you will need to water.

I know you would all like to be told when and how often, but this is the one piece of information you must find out for yourself. These are the conditions to help you to work it out.

Bonsai must be thoroughly saturated and not watered again until the soil reaches the slightly dry, not bone dry, (perhaps we should have said slightly moist) condition.

In summer it will reach this condition more quickly than winter.

The more sun it receives, the more windy conditions, even in winter, will help it to quickly become dry. The difference between receiving morning or afternoon sun also varies its drying out rate.

The more thorough the saturation, the longer the period between watering. Light sprinklings are dangerous. The water does not penetrate deep enough to supply the feeder roots at the bottom of the pot, so they die back. New feeder roots which have developed close to the surface encouraged by the water, are very vulnerable as they can dry out too quickly.

This point may be of help to you. Wet soil is dark in colour, dry soil is lighter. When the surface of the soil is looking light, dig a little into the soil and if it is darker, indicates moisture is present. This fluctuation between sopping wet and slight dampness is the healthiest condition you can give the root system of your tree.

chapter 18

Repotting

After a year or two, the plant has become rootbound, and in order to keep it HEALTHY about ⅓ of the roots must be removed from the bottom and sides of the rootball, and the space filled with your soil mix.

Allowing the Bonsai to remain in a rootbound condition for many years, causes the plant to deteriorate. Its growth slows down and becomes weaker. Now, this is what many people think Bonsai is; A STUNTED TREE.

Fig. 32   *Thickness of trunk in relation to height of tree*

Consider what is one of its most attractive features, and the answer will be the thickness of the trunk in relation to the height of the tree.

The quicker you grow your tree, and this is achieved by annual repotting, the more often you will have to top prune, but this increased growth is the means of thickening the trunk and branches.

Repotting is done at the correct time for the species as indicated in the root-pruning and potting section.

chapter 19

Pruning and trimming

Whilst still developing girth and size to your Bonsai, pruning rather than trimming will aid in achieving this goal at a faster rate. What happens is that the longer the shoots extend before being cut back the quicker the thickening process. During this period you cannot expect your Bonsai to look tidy, as the reduction of long, strong branches imparts an unrefined look to the plant. When you feel there is sufficient size, then constant trimming of the shoots develops a finer more delicate appearance. It is important to remember the more diligent your trimming with a young plant, the longer it takes to increase the framework, as you are retarding its growth.

With pruning you reduce the branch or trunk to the required length. With repeated trimming the twig is cut back to the first leaf. Each time you trim, another twig develops.

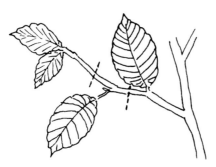

Fig. 33   *Trim back to 1st or 2nd leaf*

chapter 20

Fertilizing

Even though your soil mixture is moderately rich in humus, additional food must be supplied to maintain plant vigour. In the tropics the trees are growing at a rapid rate and combined with more frequent watering the soil leaches its nutrients very quickly.

Fertilizer should not be applied until at least six weeks have elapsed after repotting. This gives time for the roots to regenerate, thus eliminating the risk of them being burned before the cut roots have healed.

Basic fertilizers consist, in varying degrees of NITROGEN, PHOSPHORUS and POTASSIUM and to a very lesser degree — TRACE ELEMENTS.

A young vigorous plant, receiving a high nitrogen fertilizer, will grow and expand at an even more rapid rate. It will of course, encourage large lush leaf growth as well, but in the early stages this is an advantage for it all helps to thicken trunk and branches. Superphosphate added to the soil mix at the time of repotting (an exception to the rule will be Australian and some New Zealand native plants) will aid the root-system, the flowers and the fruit. The benefits for Bonsai culture means the roots are strengthened, the spaces between the leaves are shorter, the leaves are smaller, more profuse flowering and more abundant fruiting.

The third element is POTASH on which plants rely for their general well-being. Choose a fertilizer for the particular purpose needed. The ratio is denoted by three numbers. Say for instance, 5:10:8. This would mean the fertilizer would aid woodiness, fruiting and flowering, as nitrogen, the growing element is the lowest in this formulation.

If you have a problem reducing leaf size then apply a fertilizer high in phosphate during the active season. If you can obtain an 0:10:10 formula and water it in every week in areas of heavy rainfall, you should see the difference the first season. You would need to change to a higher nitrogen formula during the less active periods which preferably would be of organic origin, such as fish emulsion if obtainable. In periods of dormancy or monsoonal rains, it is not necessary to fertilize.

chapter 21

Pests and Diseases

Surprising how many people spray insecticides at half strength because the bugs are on a Bonsai. Whether pests or diseases are on plants in the garden or on a Bonsai, only the correct strength will eradicate them.

Because of the warm and often humid conditions of the tropics, pests and diseases abound. In the cooler parts of the world their succession is controlled by the severity of the winter.

For most pests and diseases the tropical climates provide ideal conditions for rapid multiplication and spread of the problems. The most common pests that attack Bonsai are aphids and caterpillars. Pyrethrum based sprays work well and are less toxic to humans. Red spider, a minute sucking mite, appears to be more troublesome in the drier climates. This is harder to eradicate than the previous pests and the remedy is to use the chemical dicofol. One trade name is Kelthane.

Mildews, both powdery and downy are a problem in the warm, humid tropical countries. They can also appear in the dry tropical regions if the plants are watered at night. The symptoms are patches of powdery like substance on the leaves. One way to tell them apart is in humid conditions a fluffy growth appears under the leaves if downy mildew is present. The chemical zineb is used for downy mildew and benomyl for powdery mildew.

chapter 22

Shaping by wire

Another popular misconception about Bonsai is that by applying wire spirally around the trunk and branches, you can stop it from growing.

Having read the chapters, guidelines for shaping and styles, you will either want to change the direction of the trunk to make it conform to one of the styles, or, to help the tree acquire the look of age the guidelines suggest certain changes in shape which will be easier to implement by wiring.

To style the trunk, insert the correct gauge wire into the soil at the back of the trunk and proceed to wire in a spiral fashion, ideally it should be wound at 45° angles.

If you have underestimated the gauge, and the trunk will not hold the shape, another wire spiralled parallel to the first should be strong enough. This point also applies to branch wiring.

Try to use one length of wire for two branches.

If wiring only one branch, it is better, for stability, to attach the wire to the trunk above and come down to the branch.

To understand the length of time needed to 'set' the wired section, varies. Brittle wood i.e. wood that cracks easily is very hard to acutely bend, but holds the position very quickly. On the other hand, supple wood on Figs and Junipers, are so easy to bend, but take such a long time to hold the shape. The best advice for the beginner is to watch the wired area and when it looks as if the wire is becoming too tight, it must be removed either by unwinding, or cutting the spirals.

If the branch springs back to its original shape, it must be rewired. The best time to wire is in autumn, when the leaves are falling. You have a clearer picture of the structure of your tree, plus, a longer time for it to spend on the tree without increased growth. By spring it can be removed.

Before you start to wire your first tree, take some branches from the garden, and practise the wiring techniques, at the same time, get the 'feel' of the various gauges.

Fig. 34   *Wiring trunk*

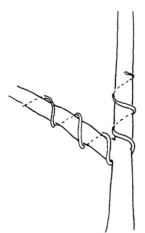

Fig. 35(a)   *Wiring single branch*

Fig. 35(b)   *Wiring branchlets*

33

chapter 23

Shaping

INTRODUCTION. AS the main difference between a potted tree and Bonsai lies in its shape, this topic represents the most important aspect of Bonsai culture. The previous chapters are concerned with the horticultural techniques, which can be applied with equal success to any tree grown in a container. This, then is the Art of Bonsai, also known as living sculpture. It is also known as the highest form of horticulture. Combine the two and you have — BONSAI.

Fig. 36(a)   *Unshaped tree is a potted tree*

Fig. 36(b)   *Excess branches removed
ready for shaping into Bonsai*

Fig. 36(c)   *Wired and shaped
skeleton of future Bonsai*

chapter 24

Problems of shaping

I suppose every Bonsai grower approaches this problem in different ways. Some start wiring their tree hoping that some 'style' will emerge in the end. Others, after seeing a picture of a Bonsai start shaping their tree regardless of the original shape of their specimen . . . some . . . sit and contemplate . . . Some try to learn the guidelines which will give some direction to follow and aim towards.

No tree when obtained will fit any of the following styles perfectly, but at the same time — almost every tree will be close in form to one of them. If you bought, for example, a tree with a horizontal trunk, do not jump to the conclusion that a semi-cascade is the best style for this tree. Consider first if a full cascade might not be the best, or if planted at a different angle, may become a better slanting style.

Fig. 37   *Not necessary to use full height of stock*

Do not feel that the full length of the plant must be used. As long as there is a suitable point where the trunk can be shortened without giving a cut off appearance it acts as an illusion that the trunk is thicker than it actually is.

chapter 25

Guidelines for shaping

The question — How does one begin to learn how to shape? is hard to answer, as shaping is an Art. Anyone can be taught art, but very few become Artists. However, don't let that statement discourage you. The following guidelines apply to all the five basic styles.

1. The trunk should taper from the base of the trunk to the apex. Any reduction in the height of the trunk must allow for a new leader to continue the taper.

2. The lower branches must be the heaviest, the rest should become thinner as they progress up the tree.

3. The branches should be distributed in groups of three, a right or left branch, a back branch, and again a left or right branch. A gap is left between each group as they ascend the trunk.

4. The back branches are the shortest, and are very necessary to provide depth and perspective.

5. Foliage close to the trunk should be removed from the branches.

6. Trim any foliage growing underneath the branches.

7. The roots should radiate out from the trunk, and no root should project towards the front.

8. The main roots should lie along the top of the soil, claw shaped roots with space inbetween, are unsightly, and don't look natural.

9. The placement for the lower branches is roughly ⅓ of the total height of the tree.

10. Two thirds up the trunk a short branch can grow towards the front.

11. The Formal upright trunk must be straight and vertical.

12. In all other styles the apex or tip inclines towards the viewer.

13. The overall shape should be triangular, with either an acute apex or a softly rounded one. Avoid flat tops.

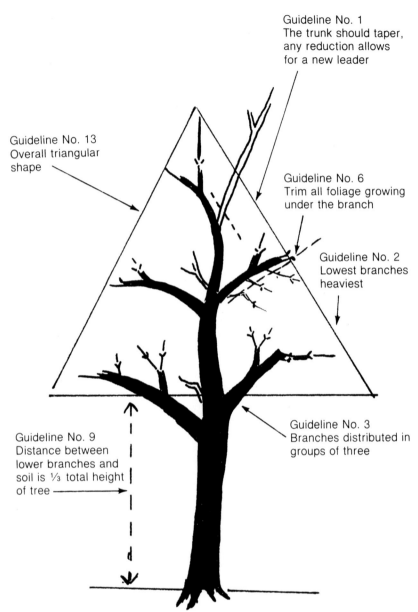

Guideline No. 1
The trunk should taper,
any reduction allows
for a new leader

Guideline No. 13
Overall triangular
shape

Guideline No. 6
Trim all foliage growing
under the branch

Guideline No. 2
Lowest branches
heaviest

Guideline No. 9
Distance between
lower branches and
soil is ⅓ total height
of tree

Guideline No. 3
Branches distributed in
groups of three

Fig. 38(a)   *Guidelines for shaping*

14. Sub branchlets should be developed, wired horizontally, in an alternate fashion, and then trimmed to an egg or pear shape with the widest end closest to the trunk.

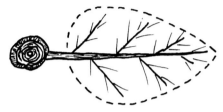

Fig. 38(b)   *Guideline No. 14*

15. No side branches should be coming from the same level. This will give the appearance of a bar across the front of the tree. A side branch and a back branch can grow from the same level as they can't be seen.

Fig. 38(c)   *Guideline No. 15*

When we first learn to write, we slavishly follow the rules and copy the perfect examples. As our hand becomes firmer, and we gain experience, we develop our own style. So it is with Bonsai. Don't be afraid to copy to start with, once you can comprehend the reasons for the rules, your breaking of them is what will add the 'character' that distinguishes your tree from all others.

There are no shortcuts here. You must learn the rules before you can break them.

chapter 26

Styles

There are five main styles, and these are determined by the position of the apex of the tree. The guidelines in the previous chapter will help you to style them.

1. Formal Upright.
2. Informal Upright.
3. Slanting style.
4. Semi-cascade.
5. Full cascade.

Fig. 39   *Formal upright*

Fig. 40   *Informal upright*

Fig. 41   *Slanting*

41

Fig. 42(a)   *Semi cascade, the tail is about level with rim of pot*

Fig. 42(b)   *Semi cascade, the tail is below rim of pot but not below base*

Fig. 43 *Full cascade*

There are many other styles, and the guidelines also apply, but we have included some additional information for each.

- Twin Trunk. The main trunk is the thickest and the most vertical. There should be no branches below the crotch.

Fig. 44   *Twin trunk, similar rules apply to two trees*

- Windswept. The branches are all shaped to one side as if windblown. The trunk can either lean in the direction of the branches, or opposite them.
- Literati. Has a very tall elegant trunk with 'character' and sparse branches and foliage starting ⅔ up the tree.
- Two tree. The main tree is the tallest, thickest and most vertical.

Fig. 45   *Literati*

44

- Group Plantings, always have odd amounts of trees. No. 1 tree is tallest and thickest, the remaining trees are graded down to the shortest and thinnest.

- Root over Rock. The tree is planted on the rock and the roots are placed along any grooves present, making sure they closely follow the contour. The roots should reach the soil below the rock.

- Clinging to Rock. The tree is planted entirely on the rock. Wires are glued to the rock to support it. The 'soil' used is called 'muck' which consists of 50% clay, 50% cow manure. Enough water is added to reach a putty like consistency. This is pressed into the root system and then covered with moss.

- Saikei, pronounced sigh-kay, means Living Landscapes. The placement of rocks and trees are of equal importance. A vast subject in itself. Its creator, Mr Toshio Kawamoto's book is recommended.

Fig. 46   *Clinging to rock*

If you are artistically inclined and have an eye for design, you will make a creditable effort. Even so, studying Bonsai which have been grown by the experts or by appraising the pictures of Bonsai in Japanese books will help you to recognise the necessity for giving your tree certain little alterations, which will make your young tree look much older, and not like the sapling it originally was.

45

You can study the old trees of the forest, and notice that many of them have their main root system exposed above the ground. Such trees look older and have added character.

You will notice too, that the lowest branches are always the heaviest, starting some distance from the ground. This too distinguishes a tree from a bush.

Looking upwards, you will see the main branches springing out from the trunk without any small sub-branches growing on them close to their base. This also is a characteristic of an older tree.

Looking at trees from a distance you will notice that the trees that attract you are not those with dense masses of foliage.

If you look at such trees, you will find that there is more airspace than branches and foliage.

The inclination of the branches is an indication of the age of the tree. Young trees still have branches growing upwards, while in middle age, they tend to grow more horizontally. As the tree reaches its maturity, the branches assume a downward position, the lowest drooping more, as they have been longer on the tree.

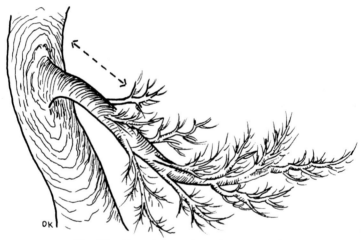

Fig. 47   *No sub-branches close to trunk*

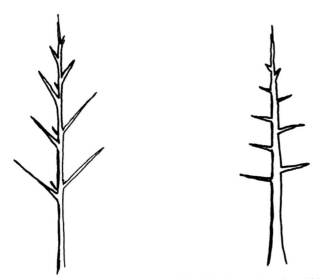

Fig. 48 *Branches on young tree*   Fig. 49 *Branches in middle age*

Fig. 50 *Mature branches*

Notice these points and by applying them to your trees you will be able to give an aged look to your Bonsai, which, after all is the goal we are aiming for.

chapter 27

Some suitable varieties for beginners

Adenum obesum.
Bougainvillea.
Callistemon. (Bottlebrush).
Casuarina. (Beef Wood, Ironwood, She Oak, Australian Pine).
Duranta. (Golden Dewdrop).
Eugenia.
Ficus sp. (Fig).
Gardenia radicans.
Grevillea robusta. (Silky Oak).
Juniperus sp. (Juniper).
Lagerstroemia. (Crepe Myrtle, 100 Day Flower).
Lantana.
Murraya.
Podocarpus.
Punica. (Pomegranate).
Serissa foetida.

In conclusion

We hope this book has whet your appetite to start Bonsai. If you only want a few trees to occasionally bring inside, this book will suffice but, if the Bonsai bug has bitten, and you want to learn more, may we recommend the most detailed and comprehensive book yet written on Bonsai.

'Bonsai Its Art, Science, History And Philosophy', by our daughter Deborah R. Koreshoff. The title speaks for itself.